iScience
Readers

Levers and Pulleys:
Lift Anything!

by Emily Sohn and Frederick Fellows

Chief Content Consultant
Edward Rock
Associate Executive Director, National Science Teachers Association

NORWOOD HOUSE PRESS
Chicago, Illinois

Norwood House Press
PO Box 316598
Chicago, IL 60631

For information regarding Norwood House Press, please visit our website at
www.norwoodhousepress.com or call 866-565-2900.

Special thanks to: Amanda Jones, Amy Karasick, Alanna Mertens, Terrence Young, Jr.

Editors: Barbara J. Foster, Diane Hinckley
Designer: Daniel M. Greene
Production Management: Victory Productions, Inc.

Paperback ISBN: 978-1-60357-292-7

The Library of Congress has cataloged the original hardcover edition with the following
call number: 2010045061

Printed in Heshan City, Guangdong, China.
190P—082011.

CONTENTS

Note to Caregivers:

Throughout this book, many questions are posed to the reader. Some are open-ended and ask what the reader thinks. Discuss these questions with your child and guide him or her in thinking through the possible answers and outcomes. There are also questions posed which have a specific answer. Encourage your child to read through the text to determine the correct answer. Most importantly, encourage answers grounded in reality while also allowing imaginations to soar. Information to help support you as you share the book with your child is provided in the back in the **Additional Notes** section.

Words that are **bolded** are defined in the glossary in the back of the book.

Simple Machines Can Do Big Jobs

Vending machines. Dishwashers. Dump trucks. Bicycles. Machines are all around us. We use them to make our lives easier.

In this book you will learn about two simple machines. You will also solve a puzzle: How can you get a 700-pound piano to the second story of a building? Sound impossible? Keep reading. It might be easier than you think!

Can You Lift 700 Pounds?

You own a moving company. You have been hired to move a grand piano. Here's the problem: You have to get it up to the second floor. The piano weighs about 700 pounds (317.5 kilograms). It's more than 6 feet (1.8 meters) deep, almost 5 feet (1.5 meters) wide, and more than 3 feet (0.9 meter) high. There is a wide staircase with 25 steps. There is also a 7 foot by 8 foot (2.1 meter by 2.4 meter) window into the front of the building. It is 15 feet (4.6 meters) from the ground. How can you use simple machines to get the piano where it needs to go?

Idea 1: Put the piano on one end of a huge seesaw. Then, throw something heavy on the other end of the seesaw. The piano will fly into the window. You can guide it in with a rope.

Idea 2: Put the piano on wheels. Cover the stairs with flat planks of wood. Then, push the piano up the planks.

Idea 3: Tie a rope around the piano. Throw the other end into the window. From the other side of the window, pull on the rope and coil it like a spool of thread. This will pull the piano in through the window.

Idea 4: Use a few of these strategies together.

To solve the puzzle, you will need to learn about two kinds of simple machines—the lever and the pulley. You will have to figure out how to use the machines correctly. And you may need to be creative: Is there a way you could use both machines together to solve the puzzle? Put on your thinking cap and keep reading!

Do you think this would be a good way to lift a piano?

Balancing Pennies

Have you ever been on a seesaw? A seesaw works best if you and your partner weigh about the same. To see why, make a model with a pencil and a flat ruler. Balance the ruler by laying the center of it across the pencil. Place a single penny on one end of the ruler. Place a stack of five pennies on the other end. The heavier end of the seesaw will drop. Game over? Not yet. You may be able to make the ruler balance again.

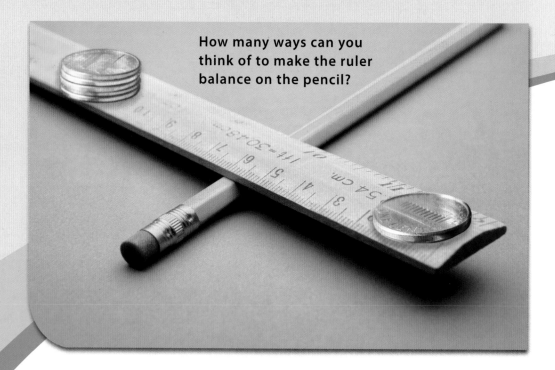

How many ways can you think of to make the ruler balance on the pencil?

What happens when you move the bigger stack of pennies toward the middle of the ruler? What else could you do to find the balance? Make a list of your ideas. Then, test them one at a time. Here's one hint: The pencil and ruler are not attached to each other. Does that give you any ideas?

Using a machine can make
your job easier!

Congratulations. You made a machine! It may not be a lawn mower or a high-speed train. Still, when you use a ruler this way, it is a type of simple machine called a **lever.** A lever is a hard bar that sits on one point, called a **pivot point.** With a lever, the pivot point doesn't move. Instead, the bar can swivel around on the point.

With your machine, you lifted pennies. Some levers lift objects that are much heavier or harder to move. The back end of a hammer is just such a lever. Say you want to pull a nail out of a piece of wood. Your fingers can't budge it. Instead, you can wedge the head of the nail between the hammer's claws. Then use the hammer as a lever to pry the nail out.

A simple machine can help you lift something you could never lift on your own.

Levers can perform amazing feats. Some car jacks use a lever and gears to lift up one end of an automobile. Some jacks can pull a house off its base! Jacks are small tools. But levers give them the power to lift hundreds or even thousands of pounds.

With the ruler, you did a few experiments. You tested one idea. Then, you tested another. This is exactly what scientists do. As they work, they take careful notes. They also study their mistakes. That helps them figure out exactly what works and what doesn't.

Scientists often think about
and test lots of ideas before
they find one that works.

Making your own machines can be frustrating. You might not get it right the first time. Keep tinkering! It can be fun to build something that makes life easier.

Now, think about the piano problem. What kinds of experiments could you do to find the best way to move it? Do you think a lever might help? Read on. First, you need to learn some basics.

What Is a Force?

Any time you want to move something, you face a basic problem. Ordinary objects can't get up and walk on their own. They can't fly, either. To move the piano, you will need to use a **force.** A force is a pull or a push that makes an object move, stop moving, or change the direction in which it's moving.

Gravity is a force that earthlings know well. Gravity affects anything that has **mass.** Mass refers to the amount of matter, or "stuff," in an object. For example, a basketball has more mass than a beach ball. Objects with more mass are more affected by Earth's gravity, and this gives them more weight. Gravity is why the Moon orbits Earth. It's also why pianos sit on Earth—not the other way around.

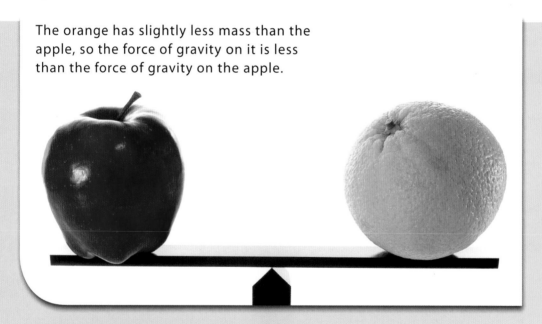

The orange has slightly less mass than the apple, so the force of gravity on it is less than the force of gravity on the apple.

Scientists measure forces in newtons. This **unit** is named after the English scientist Isaac Newton. You will read about him later. This unit describes the force needed to move a 1-kilogram (2.2-pound) object at a certain speed. After it starts moving, the object keeps picking up speed.

Right now the two teams are pulling with equal force in opposite directions.

Draw It!

So, you're scratching your head and looking at the piano. You can't get enough force with your arms to lift it. Clearly, you need to build a machine. But what should the machine look like? And how will you build it?

Scientists who build things often start with a drawing. They use arrows to illustrate forces. Look at this picture of a tug-of-war game. Each team pulls in an opposite direction from the other. The teams are evenly matched. So, the forces cancel each other out. Forces can move in any direction. The size of the arrows shows how big the forces are.

Draw a picture of the piano. Draw arrows to show the forces that are acting on it right now. You can also draw the forces needed to move it. Make sure to show how big the forces are compared to each other.

Having help is another way to lessen the effort you have to put into moving something. Two people mean twice the force.

Forces can work together. You can't lift the piano on your own. But you might succeed if you get friends to help. In your drawing of the piano, add yourself and a group of friends. Draw an up arrow next to each person. You can put longer arrows next to stronger people. Each arrow is a lifting or pushing force. If you add up all of the arrows, you will see how much power you can get from teamwork.

Galileo Galilei

Galileo and Newton

When you hear the name Galileo Galilei, you may think of outer space. Galileo was an Italian scientist who lived in the late 1500s and early 1600s. He was the first to find moons around Jupiter and spots on the Sun. He supported Nicolaus Copernicus's theory that the Sun was at the center of our solar system. At that time, people believed that everything revolved around Earth.

But Galileo's mind wasn't always in outer space! He also experimented with metal balls here on Earth. He imagined a ball rolling on a perfectly smooth surface. He realized that, with no force to slow it down, the ball would roll forever.

Isaac Newton

About 100 years later, Isaac Newton picked up Galileo's idea that a ball with no force acting on it would roll forever and experimented with it. Newton was born in 1643 in England. In his life, he came up with the three Laws of Motion. Imagine living in a time when the study of science was still brand new! Both Galileo and Newton worked hard to understand how the universe worked.

Newton's Laws of Motion can help you build all sorts of machines. The first law is often called the law of **inertia.** It explains that an object at rest will stay at rest and an object in motion will stay in motion. It takes a force to make an object move, stop moving, or change direction. In other words, that piano is just going to sit there until you use a force to make it move.

In the microgravity of space, astronauts have to make sure they don't push or pull too hard to move an object.

You may have seen videos of astronauts in outer space. Everything and everyone seems to float. If you brought a piano into space, it would float too. You could push it with your pinky finger. Then, it would move in a straight line until some force stopped it or changed its path. This situation is called microgravity. The piano would be falling, but because Earth's gravity far out in space is so weak, the piano would fall around Earth instead of into Earth. The piano would be in orbit!

It's too bad you are moving a piano on Earth, not in space! Still, Newton's laws may yet help you set the piano in motion.

Moving an object with a little mass is much easier than moving an object with a lot of mass.

Shrink It!

You've already drawn the forces that are acting on your piano. Another way to solve your piano problem is to shrink it. Draw two pianos. Make one a grand piano that weighs 700 pounds. Make the other one a toy piano that weighs 2 pounds. Which one would be easier to move?

Newton's second law explains why you probably chose the toy. It takes more force to move an object that has more mass. Moving a tiny piano doesn't take much force at all.

Any object with wheels requires less force to be moved than does the same object without wheels.

Let It Roll!

Now draw wheels on both of the pianos. Wheels often make objects easier to move. It would still take more force to move the heavier piano. But imagine pushing both pianos with the same amount of force. In this experiment, the lighter object would accelerate, or pick up speed, more quickly.

Once the two pianos were moving, it would take more force to stop the bigger one. That's because heavier objects have more inertia. If you pushed both pianos down a ramp with the same amount of force, the bigger one would actually roll farther.

Shopping carts have wheels. So do cars and bikes. Why do you think wheels make it easier to move objects? Would it get harder or easier to push the piano if you started going up a hill?

Friction

You've learned that a moving object tends to stay in motion until a force stops it or slows it down. But here on Earth, moving objects stop all the time. You roll a ball, throw a plastic disk, or sled down a hill. None of those movements goes on forever.

Friction is the main reason why. You can think of friction as a rubbing force that slows things down and produces heat. Have you ever skateboarded or roller-skated? As your wheels roll forward, the ground pushes back on them. This creates friction. You have to keep working to keep moving. If you touch the wheels after a ride, they might feel hot. Heat is a form of **energy.**

Friction is acting on the wheels of this skateboard.

Skateboard parks are made out of smooth, slick surfaces. The smoothness reduces friction, letting skaters roll farther and faster. Low friction explains why you can slide along a hardwood floor in your socks but you can't do the same on the carpet. It also explains why you might slip on wet ice or a banana peel.

Do you think oiling the chain on a bicycle adds friction or takes it away? Would putting wheels on the piano increase or decrease friction?

How Are Force, Work, and Energy Related?

Doing homework. Mowing the lawn. Cleaning your room. Life is full of work. So is science. In science, **work** describes a force that makes something move along a distance. Your fingers do work when they write with a pencil. Your arms do work when they use a shovel. It takes more work to pick up a heavier shovel than a lighter shovel. And it takes more work to throw what's in the shovel a longer distance.

You do work when you lift or move something, such as this snow.

Think again about newtons. Remember that newtons measure forces. Joules are the units used to measure work. It takes about one joule of energy to lift an apple one meter off the ground.

Here's a simple formula:

Work (joules) = Force (newtons) × Distance (meters)

Say you use a force of 200 newtons to move a piano a distance of 10 meters. Multiply 200 by 10. That's 2,000 joules of work. How much work will it take if you and three friends each use 200 newtons to move a piano a distance of 50 meters?

When the woman releases it, the football will have kinetic energy.

Did you have breakfast this morning? Let's hope so! Food provides energy for your muscles and your brain. You're going to need both to solve the puzzle and to move that piano. Energy is the ability to do work. Airplanes and trucks get energy from oil and gas. Kids get energy from pizza, apples, and other foods.

Out in the world, energy can take on more than one form. Imagine playing catch with a football. When you throw the ball, you put energy into it. As the ball flies, it is full of **kinetic energy.** This is the energy of motion. A moving car, a skipped stone, a diving roller coaster, and a whooshing sled are all examples of objects with kinetic energy.

A roller coaster's energy changes as the coaster moves around the loop.

A roller coaster has kinetic energy as it flies down a hill. At the top of a hill, it has **potential energy.** This is energy that is stored.

There is potential energy in a light bulb that is off and a car that is sitting still. Other examples of objects or substances with potential energy are a sled sitting at the top of a hill, water behind a dam, and soda in a shaken but unopened can. What else around you is full of stored energy?

Energy and work are two different things. If you hold one end of the piano up, you are using energy. But to make the piano move, you also need to do work.

How Do Levers Affect Forces?

So, the piano is still sitting there. Its owner is waiting for you to move it. Now that you know about forces, energy, and work, let's look back at the iScience Puzzle and think about Idea 1. The idea is to use a lever to lift and hurl the piano through a big open window.

Think about the lever you made with a ruler and a pencil. The pencil, or pivot point, is called the **fulcrum.** The parts of a lever on either side of the fulcrum are the **lever arms.**

Imagine that you had a big lever with a piano on one arm. What could you put on the other arm to lift the piano? How would you get any such object onto the lever? And if you could hurl the piano through the second-story window, how would you stop it safely?

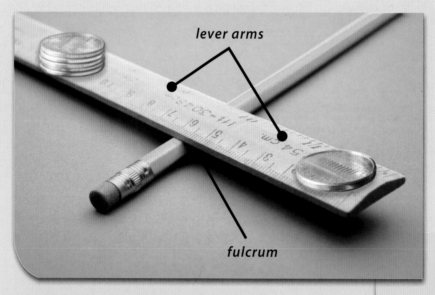

lever arms

fulcrum

These lever arms are just big enough to hold pennies. Imagine lever arms big enough to hold a piano!

approximate path of the piano through the window

To get at a yummy fruit, the orangutan can use a stick to pry open the fruit's shell.

Simple, but Full of Force

People have known about the power of levers for a long time. So have animals. In the wild, otters use rocks as levers to pry open shells. Orangutans use sticks to pry open prickly fruits.

Levers work so well because they can turn a small force into a bigger one. The force you put in is called the **input force.** What comes out is the **output force.** The difference between the two is called the **mechanical advantage** of a lever.

Earlier, your ruler used the weight of one penny to lift five pennies. That gave it a mechanical advantage of 5. You can also write this as 5:1. Think about a 700-pound piano. Compare the piano's weight to your own. Would a mechanical advantage of 5:1 be enough for you to lift the piano? Hint: Multiply your weight in pounds by 5. Is the answer more than 700?

lever arm

fulcrum

With the fulcrum so close, and a long lever arm, this can is easy to open.

❓ Did You Know?

Archimedes, a Greek mathematician and inventor, lived more than 2,000 years ago. He knew that a lever was a powerful tool. "Give me a place to stand and I can move the Earth," he said. What do you think he meant?

One trick to getting force out of a lever is to move the fulcrum. You may have figured that out already with the pencil and ruler. If you moved the fulcrum closer to the larger stack of pennies, the pennies were easier to lift.

A screwdriver is a good example of how lever power works. In the picture, the fulcrum of the lever is about 0.5 inches (1.25 centimeters) from the lid. Meanwhile, the input arm is about eight inches (20 centimeters) long.

Say you wanted to use a lever to lift your 700-pound piano. You put the piano on the output arm. You put a 200-pound weight on the input arm. Should the fulcrum be closer to the piano or closer to the weight?

Levers with a Twist

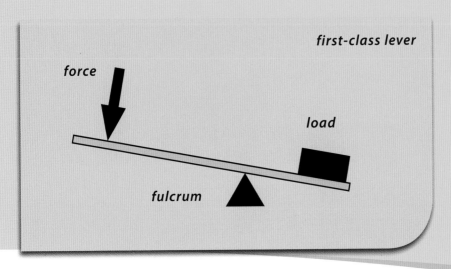

There are three types of levers. A **first-class lever** looks
like a seesaw. There are a few ways to get more force out of a
first-class lever. You can move the fulcrum. You can change the arm
length. You can put two together to make one machine. Scissors
are made of two first-class levers used together.

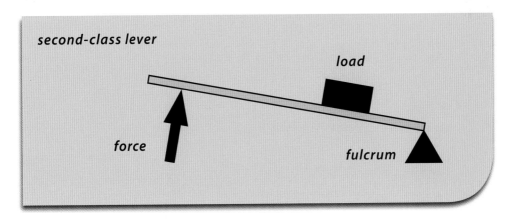

If you switch the load and the fulcrum of a first-class lever,
you make a **second-class lever.** Common examples include
wheelbarrows and bottle openers. Nail clippers and nutcrackers
are made of two second-class levers used together.

27

A lot of effort goes into hitting a baseball with a bat. But the payoff is that the ball travels a great distance (you hope!).

third-class lever

load

force

fulcrum

To lift a really heavy object with a first-class lever, you need to put the fulcrum very close to the load, like a screwdriver lifting the lid off a paint can. The problem with this kind of lever is that you can't lift the load very far. You can solve this problem by using a **third-class lever.** With a third-class lever, one end of the lever arm sits on the fulcrum. The load sits on the other end. You add force somewhere near the middle.

You have third-class levers in your body. Watch your arm as you pick up something heavy. Your shoulder is the fulcrum. The load is in your hand at the other end of the lever. The muscles in between (the force) do the work.

A baseball bat is another third-class lever. You have to put in extra effort to move this kind of lever. But the payoff is that the load (in this case a baseball) will travel farther.

Which type of lever do you think would work best for moving a 700-pound piano? Look at Idea 1 again. Which type of lever does it use?

With a wheelbarrow, you can haul heavy loads great distances.

axle

Making Levers Work for You

Wouldn't it be great if you could multiply the work you do? Say you did just one math problem. Four others would magically be done without any extra work from you! Levers do something like that. You put in a certain amount of force. The machine multiplies the work.

Think about a wheelbarrow. The axle of the wheel is the fulcrum. You lift the handles, which are the input arms. The load sits in the middle of the simple machine. This second-class lever lets you haul heavier loads than you could carry in your arms.

How do you think the stones were moved across the desert and raised to build the pyramids?

People have been building massive structures for thousands of years. The pyramids in Egypt, Stonehenge in England, the stone buildings of Machu Picchu in Peru are often considered to be great wonders of the world. They were built with stones that weigh thousands of pounds. People carried those stones for many miles. They did all of this without the help of cranes or other heavy machinery.

How do you think people moved heavy objects hundreds, or even thousands, of years ago? Do some research to find out how the structures mentioned on this page were actually built.

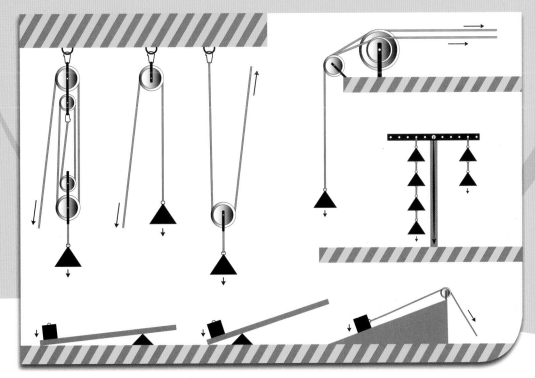

Do you see a machine in the picture that looks like it could move your piano?

To Lever or Not to Lever

You've seen how levers can lift heavy loads. Based on what you've learned, do you think a lever could help you get the piano up to the second floor? What would the drawbacks of a lever be?

A **pulley** is another kind of simple machine. Even if you've never heard of pulleys, you've probably seen or even used them. They are often used to lift heavy objects at construction sites.

A mechanic is using a wrench, a type of lever, to loosen an oil drain plug on a car.

Auto Mechanic

One of the best places to see levers in action is at an auto shop. Mechanics use a lot of tools. Some of them are complicated. To raise a 3,000-pound (1,361-kilogram) car, for example, a hydraulic lift uses an electric pump and oil under high pressure.

But many of a mechanic's tools are simple. To lift just a few hundred pounds, he might use a hand winch. This tool is also called a come-along. A winch winds up a cable like a spool of thread. A handle works as a lever, supplying the force needed to wind the cable.

Mechanics also use wrenches as levers to loosen and tighten bolts. If a small wrench can't loosen a very tight bolt, a longer wrench might work. Which class of lever is a wrench?

Pulley Systems and How They Affect Forces

Levers aren't the only machines that can give you power. You can also get help from systems that use ropes and wheels. In this section you will learn about pulleys. A pulley is a grooved wheel with a rope passing through the groove.

Look at this photo of a bucket. The pulley is attached to one spot. This is a **fixed pulley.** A pulley can also be hooked to an object that moves,

fixed pulley

With this fixed pulley, you raise and lower the bucket, but the pulley itself stays in one place.

By pulling down on the free end of the rope, you can raise the bucket with only a little effort. When the child pulls down on the rope, the bucket is lifted with an equal but opposite force.

In One Way, Out the Other

Say you want to pull a big bucket of water out of the well in the picture above. Water is heavy. Trying to just yank the bucket up with the rope would be really hard. With a pulley, you pull the other end of the rope down instead of up. This kind of pulley doesn't actually give you a mechanical advantage. But it changes the direction of your force. By yanking downward, you can really put your weight into the effort. You can raise the bucket of water. What would happen if you tried to just pull the bucket up by the rope?

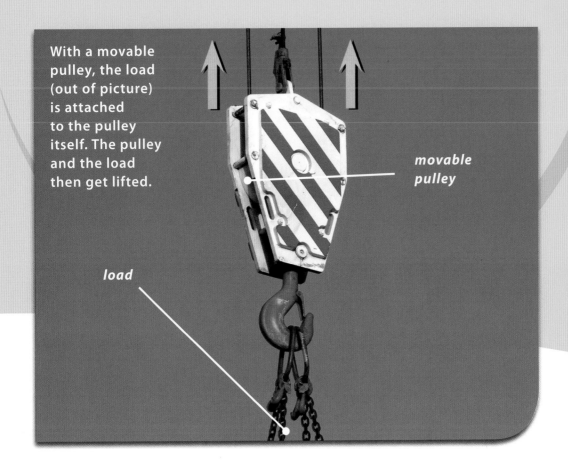

With a movable pulley, the load (out of picture) is attached to the pulley itself. The pulley and the load then get lifted.

movable pulley

load

More Bang for Your Buck

To get more power out of a pulley, you need to get it moving. With a **movable pulley,** you actually attach the pulley to the load you want to lift. First, you tie one end of the rope to something that doesn't move. The rope goes under the pulley's grooved wheel. The load to be moved hangs from the pulley by a hook. You pull on the free end of the rope that is attached to the pulley.

Because there are two ropes—one attached to the pulley, and one attaching the pulley to the object being lifted—you have to lift only half of the weight of the object. The extra rope holds the other half. That makes your job easier. If there were no friction, this kind of pulley should double the force you put in.

35

fixed pulley

movable
pulley

With pulleys, you can also mix and match. In the picture above, you see a fixed pulley above and a movable pulley below. The fixed pulley changes the direction of the force you put in. The movable pulley multiplies your force. A system of fixed and movable pulleys is called a **block and tackle.**

This kind of system makes a load easier to lift. It also makes the work take longer. That's because each yank pulls the weight a shorter distance compared to a fixed pulley.

Would it be easier to lift the piano with a fixed pulley or a movable pulley? Would you rather use just one pulley or a block and tackle system?

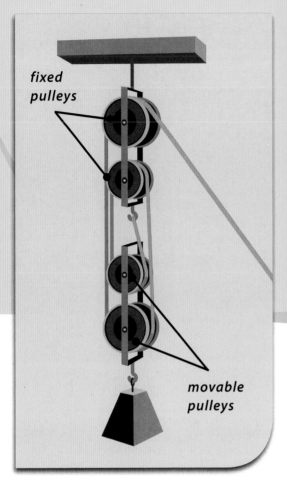

fixed pulleys

movable pulleys

Every turn of the rope in a block and tackle system adds more force.

The More the Merrier

There's no limit to the number of pulleys you can string together. Look at the block and tackle system in this illustration. It has two fixed and two movable pulleys.

The pulleys in a block and tackle system turn easily. And every rope adds the same amount of force. With more pulleys and more ropes, your work becomes even easier.

About how much force would it take you to lift this man? How would you do it?

Here's how to figure out how much force you need to put into a block and tackle system. First, write down the weight of the load you are trying to lift. Then divide that number by the number of ropes that support it.

Let's say you have a block and tackle with four ropes that support an object. Each rope carries one-fourth of the weight. That means that you have to pull with only one-fourth as much force to lift the load. Suppose you wanted to lift a 200-pound man with this system. How much force would you need to use?

You can think about this the opposite way. With four ropes, you get out four times as much force as you put in. So, the mechanical advantage is four.

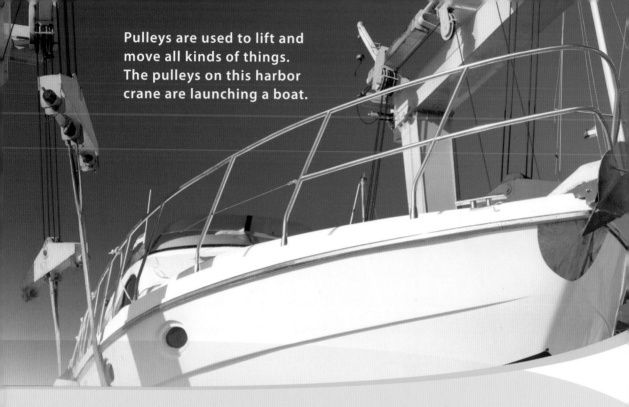

Pulleys are used to lift and move all kinds of things. The pulleys on this harbor crane are launching a boat.

Think about the iScience Puzzle again. The piano weighs 700 pounds. Say you have a system with five pulleys. How much force would you need to put in to lift it?

How many pulleys would the system need to have if you wanted to lift the piano by yourself? How many pulleys would you need if you had help from a few friends?

Pulleys have many uses on boats. They lift cargo and fishing nets, and they raise and lower sails.

It's time to move! The owner of the piano has a concert coming up. She needs to practice. And she wants the piano upstairs—now! You've learned about levers. And you've learned about pulleys. How are you going to decide which to use? First, consider what they have in common.

Both are simple machines. Both multiply the amount of force you put into them. Both can lift hundreds of pounds or more. Both also add distance between you and the load. To lift a piano with a first-class lever, you would need a really long lever arm on the input side. With a pulley, the distance is taken up by coiled rope. Which machine would take up less space?

This boat is being moved on a sling.

Levers work better than pulleys for some kinds of work. For example, it probably wouldn't make sense to use a pulley to open a can of paint or to staple two pieces of paper together. But pulleys are best for other kinds of work. They're a good choice when you need to lift something up really high.

Levers and pulleys can also work together. Remember the winch at the auto shop? This tool uses a lever to get the force needed to wind a cable around a pulley. The device allows you to wind the lever in only one direction. It holds the cable in place until someone releases it. Cranes use winches and pulleys to lift heavy objects for construction projects.

When the object you want to move is very large, a sling can also help. Slings spread weight across a bigger area. That makes the object less likely to break or fall. Do you think you could use a sling to move a piano?

moving an upright piano along a ramp

You're ready to move the piano. The owner is watching. Which method did you choose? Let's consider each idea.

Idea 1: Use a first-class lever to hurl the piano through the window. This might work. But you would need to build a huge lever. And the piano would most likely get damaged. That would not make the owner happy.

Idea 2: Push the piano up a ramp on wheels. The wheels would reduce friction and make the piano easier to push. But gravity is a strong force. The piano might come rolling back down on you. A runaway piano is not a good thing. Taking a grand piano to the second story on a long ramp would be risky indeed!

Idea 3: Pull the piano through the window, using the windowsill and a rope as a fixed pulley. A pulley is a helpful and safe way to lift a piano. But a fixed pulley doesn't multiply the force you put in. It only reverses the direction of the force. It might be too hard to pull the piano up this way.

Idea 4: Combine strategies. The best way to move the piano is with both levers and pulleys. A block and tackle system combined with a winch or two would probably do the trick. You could even use a crane.

Pull out a pencil and piece of paper. See if you can draw the perfect system for moving the piano. Explain the mechanical advantages you'd get. Show your math to prove that your system can lift 700 pounds (317.5 kilograms). Then, sit back and enjoy the piano music!

Movable cranes are often used to move construction materials. But they can move all kinds of things—even pianos!

You've learned a lot about how levers and pulleys can make life easier. Both are types of simple machines. There are six kinds of simple machines. A ramp, like the one mentioned in idea 2 of the iScience puzzle, is another example. You may have seen these simple slopes behind a moving truck. It is easier to carry boxes up a ramp than to lift them straight up from the ground directly onto the truck. In science, a ramp is called an inclined plane.

A fourth kind of simple machine is a wedge. This triangle-shaped machine can pull objects apart or lift them up.

The fifth type of simple machine is the screw, which is an inclined plane wrapped in a spiral around an axis. You may have used a screw to fasten the battery cover on an electric toy or you may have seen one in the form of a jar lid.

inclined plane

An axe is a wedge.

The sixth type of simple machine is the wheel and axle. Remember the discussion about first-class levers? The wheel and axle is a version of one. The wheel rotates in a circle around the center point on an axle. You've likely used bicycle wheels to travel from place to place or even ridden on a Ferris wheel. But did you know that a doorknob is also a kind of wheel and axle?

Many complex machines are simply combinations of these basic devices. How many machines can you find that combine some of these devices? How many ways can you think of to use the six simple machines?

Now that you know how simple machines work, you might invent the next great machine!

wheel and axle

screws

GLOSSARY

block and tackle: a system of fixed and movable pulleys.

energy: the ability to do work.

first-class lever: a lever with the fulcrum between the load and the effort force.

fixed pulley: a pulley attached overhead.

force: a push or a pull.

friction: force between two surfaces moving past each other. Friction slows down motion.

fulcrum: unmoving point around which a lever pivots.

inertia: the tendency of an object to keep moving or to sit still in the absence of any force.

input force: force applied to a lever.

kinetic energy: energy of motion.

lever: a bar that pivots (turns) around an unmoving point.

lever arms: parts of a lever on either side of the fulcrum.

mass: the amount of matter in an object.

mechanical advantage: ratio of the output force to the input force.

movable pulley: pulley attached to the object that it moves.

output force: force produced by a lever.

pivot point: a central point on which something turns, or pivots.

potential energy: energy that is stored.

pulley: a grooved wheel with a rope that passes under it.

second-class lever: a lever with the load between the fulcrum and the effort force, such as a wheelbarrow.

third-class lever: a lever with the effort force between the fulcrum and the load, such as a lacrosse stick or a baseball bat.

unit: a fixed amount used as a standard of measurement.

work: force acting along a distance.

FURTHER READING

Super Cool Science Experiments: Levers and Pulleys, Dana Meachen Rau. Cherry Lake Publishing, 2010.

Exploratorium. **Lever Arm.** http://isaac.exploratorium.edu/~pauld/ workshops/bodymechanics/leverarm.htm

MIKIDS.com. **Simple Machines.** http://www.mikids.com/Smachines.htm

ADDITIONAL NOTES

The page references below provide answers to questions asked throughout the book. Questions whose answers will vary are not addressed.

Page 8: The ruler would balance.

Page 19: Wheels reduce friction on the object being moved. With wheels, you wouldn't have to push as hard. It would be harder to push the piano uphill.

Page 20: Oiling the bike chain would reduce friction. Wheels would decrease the friction on the piano.

Page 21: It will take 40,000 joules.

Page 25: No, 5:1 would not be a great enough mechanical advantage.

Page 26: Closer to the piano. **Did You Know?** Archimedes meant that with a large enough lever a single person could move a whole planet.

Page 28: Idea 1 uses a first-class lever.

Page 32: A wrench is a second-class lever.

Page 34: The bucket would be too heavy to lift.

Page 36: It would be easier to lift the piano with a movable pulley. A block and tackle system would be even easier.

Page 38: You would need to use 50 pounds of force.

Page 39: You would need to use 140 pounds of force.

Page 40: A pulley would take up less space.